THIS BOOK BELONGS TO

EMAIL: _____

ADDRESS: _____

CONTACT: _____

PHONE: _____

START DATE	END DATE

MO	TU	WE	TH	FR	SA	SU		DATE:	/	/
☐	☐	☐	☐	☐	☐	☐				

PROJECT: _____ **FOREMAN:** _____

WEATHER

F°_____ C°_____ _____ AM _____ PM

HOURS DUE TO BAD WEATHER	ISSUED AND DELAYS

NOTE: _____

COMPLETION DATE	DAYS AHEAD OF SCHEDULE	DAYS BEHIND SCHEDULE

SAFETY AND INCIDENTS

SAFETY ISSUES THAT NEED TO BE ADDRESSED	ACCIDENTS / INCIDENTS / STEPS NEEDED TO RESOLVE
_____	_____
_____	_____
_____	_____
_____	_____

SUMMARY OF THE WORK DONE TODAY

_____
_____
_____
_____
_____

IMPORTANT NOTES

_____
_____
_____
_____
_____

NAME	SIGNATURE

TODAY LABOR

INITIALS	TRADE	START	FINISH	PAID HOURS	OVERTIME	COMPANY
☐ EMPLOYEE ☐ CONTRUCTOR		AM	PM			
☐ EMPLOYEE ☐ CONTRUCTOR		AM	PM			
☐ EMPLOYEE ☐ CONTRUCTOR		AM	PM			
☐ EMPLOYEE ☐ CONTRUCTOR		AM	PM			
☐ EMPLOYEE ☐ CONTRUCTOR		AM	PM			
☐ EMPLOYEE ☐ CONTRUCTOR		AM	PM			
☐ EMPLOYEE ☐ CONTRUCTOR		AM	PM			
☐ EMPLOYEE ☐ CONTRUCTOR		AM	PM			

EQUIPMENT ON SITE	NO. OF UNITE	WORKING YES / NO

HIRED EQUIPMENT	NO. OF UNITE	EQUIPMENT RENTED	FROM	RATE

NAME: _____   SIGNATURE: _____

| MO | TU | WE | TH | FR | SA | SU |
| ☐ | ☐ | ☐ | ☐ | ☐ | ☐ | ☐ |

DATE: ____ / ____ / ____

PROJECT: _____

FOREMAN: _____

WEATHER

F° ____ C° ____ ____ AM ____ PM

HOURS DUE TO BAD WEATHER	ISSUED AND DELAYS

NOTE: _____

COMPLETION DATE	DAYS AHEAD OF SCHEDULE	DAYS BEHIND SCHEDULE

SAFETY AND INCIDENTS

SAFETY ISSUES THAT NEED TO BE ADDRESSED	ACCIDENTS / INCIDENTS / STEPS NEEDED TO RESOLVE

SUMMARY OF THE WORK DONE TODAY

IMPORTANT NOTES

NAME	SIGNATURE

TODAY LABOR

INITIALS	TRADE	START	FINISH	PAID HOURS	OVERTIME	COMPANY
☐ EMPLOYEE ☐ CONTRUCTOR		AM	PM			
☐ EMPLOYEE ☐ CONTRUCTOR		AM	PM			
☐ EMPLOYEE ☐ CONTRUCTOR		AM	PM			
☐ EMPLOYEE ☐ CONTRUCTOR		AM	PM			
☐ EMPLOYEE ☐ CONTRUCTOR		AM	PM			
☐ EMPLOYEE ☐ CONTRUCTOR		AM	PM			
☐ EMPLOYEE ☐ CONTRUCTOR		AM	PM			
☐ EMPLOYEE ☐ CONTRUCTOR		AM	PM			

EQUIPMENT ON SITE	NO. OF UNITE	WORKING YES / NO

HIRED EQUIPMENT	NO. OF UNITE	EQUIPMENT RENTED	FROM	RATE

NAME: _____ SIGNATURE: _____

MO   TU   WE   TH   FR   SA   SU
☐   ☐   ☐   ☐   ☐   ☐   ☐

DATE: ___ / ___ / ___

PROJECT: _____

FOREMAN: _____

WEATHER

F°____ C°____ ____AM ____PM

HOURS DUE TO BAD WEATHER	ISSUED AND DELAYS
	_____

NOTE: _____

COMPLETION DATE	DAYS AHEAD OF SCHEDULE	DAYS BEHIND SCHEDULE

SAFETY AND INCIDENTS

SAFETY ISSUES THAT NEED TO BE ADDRESSED	ACCIDENTS / INCIDENTS / STEPS NEEDED TO RESOLVE

SUMMARY OF THE WORK DONE TODAY

_____
_____
_____
_____
_____

IMPORTANT NOTES

_____
_____
_____
_____
_____

NAME	SIGNATURE

TODAY LABOR

INITIALS	TRADE	START	FINISH	PAID HOURS	OVERTIME	COMPANY
☐ EMPLOYEE ☐ CONTRUCTOR		AM	PM			
☐ EMPLOYEE ☐ CONTRUCTOR		AM	PM			
☐ EMPLOYEE ☐ CONTRUCTOR		AM	PM			
☐ EMPLOYEE ☐ CONTRUCTOR		AM	PM			
☐ EMPLOYEE ☐ CONTRUCTOR		AM	PM			
☐ EMPLOYEE ☐ CONTRUCTOR		AM	PM			
☐ EMPLOYEE ☐ CONTRUCTOR		AM	PM			
☐ EMPLOYEE ☐ CONTRUCTOR		AM	PM			

EQUIPMENT ON SITE	NO. OF UNITE	WORKING YES / NO

HIRED EQUIPMENT	NO. OF UNITE	EQUIPMENT RENTED	FROM	RATE

NAME: _____   SIGNATURE: _____

MO TU WE TH FR SA SU
☐   ☐   ☐   ☐   ☐   ☐   ☐

DATE: ____ / ____ / ____

PROJECT: _____

FOREMAN: _____

WEATHER ☁ ⛅ ☁ 🌨 ☀ 🌧 ⛈

F°____ C°____ ____AM ____PM

HOURS DUE TO BAD WEATHER	ISSUED AND DELAYS

NOTE: _____

COMPLETION DATE	DAYS AHEAD OF SCHEDULE	DAYS BEHIND SCHEDULE

SAFETY AND INCIDENTS

SAFETY ISSUES THAT NEED TO BE ADDRESSED	ACCIDENTS / INCIDENTS / STEPS NEEDED TO RESOLVE

SUMMARY OF THE WORK DONE TODAY

_____
_____
_____
_____
_____

IMPORTANT NOTES

_____
_____
_____
_____
_____

NAME	SIGNATURE

TODAY LABOR

INITIALS	TRADE	START	FINISH	PAID HOURS	OVERTIME	COMPANY
☐ EMPLOYEE ☐ CONTRUCTOR		AM	PM			
☐ EMPLOYEE ☐ CONTRUCTOR		AM	PM			
☐ EMPLOYEE ☐ CONTRUCTOR		AM	PM			
☐ EMPLOYEE ☐ CONTRUCTOR		AM	PM			
☐ EMPLOYEE ☐ CONTRUCTOR		AM	PM			
☐ EMPLOYEE ☐ CONTRUCTOR		AM	PM			
☐ EMPLOYEE ☐ CONTRUCTOR		AM	PM			
☐ EMPLOYEE ☐ CONTRUCTOR		AM	PM			

EQUIPMENT ON SITE	NO. OF UNITE	WORKING YES / NO

HIRED EQUIPMENT	NO. OF UNITE	EQUIPMENT RENTED	FROM	RATE

NAME: _____   SIGNATURE: _____

MO TU WE TH FR SA SU
☐   ☐   ☐   ☐   ☐   ☐   ☐

DATE: ___ / ___ / ___

PROJECT:

FOREMAN:

WEATHER

F°_____ C°_____ _____AM _____PM

| HOURS DUE TO BAD WEATHER | ISSUED AND DELAYS |

NOTE: _____

COMPLETION DATE	DAYS AHEAD OF SCHEDULE	DAYS BEHIND SCHEDULE

SAFETY AND INCIDENTS

SAFETY ISSUES THAT NEED TO BE ADDRESSED	ACCIDENTS / INCIDENTS / STEPS NEEDED TO RESOLVE

SUMMARY OF THE WORK DONE TODAY

IMPORTANT NOTES

NAME	SIGNATURE

TODAY LABOR

INITIALS	TRADE	START	FINISH	PAID HOURS	OVERTIME	COMPANY
☐ EMPLOYEE ☐ CONTRUCTOR		AM	PM			
☐ EMPLOYEE ☐ CONTRUCTOR		AM	PM			
☐ EMPLOYEE ☐ CONTRUCTOR		AM	PM			
☐ EMPLOYEE ☐ CONTRUCTOR		AM	PM			
☐ EMPLOYEE ☐ CONTRUCTOR		AM	PM			
☐ EMPLOYEE ☐ CONTRUCTOR		AM	PM			
☐ EMPLOYEE ☐ CONTRUCTOR		AM	PM			
☐ EMPLOYEE ☐ CONTRUCTOR		AM	PM			

EQUIPMENT ON SITE	NO. OF UNITE	WORKING YES / NO

HIRED EQUIPMENT	NO. OF UNITE	EQUIPMENT RENTED	FROM	RATE

NAME: _____   SIGNATURE: _____

MO TU WE TH FR SA SU
☐   ☐   ☐   ☐   ☐   ☐   ☐

DATE: /   /

PROJECT:

FOREMAN:

WEATHER

F°____ C°____   ____AM ____PM

| HOURS DUE TO BAD WEATHER | ISSUED AND DELAYS |

NOTE: _____

COMPLETION DATE	DAYS AHEAD OF SCHEDULE	DAYS BEHIND SCHEDULE

SAFETY AND INCIDENTS

SAFETY ISSUES THAT NEED TO BE ADDRESSED	ACCIDENTS / INCIDENTS / STEPS NEEDED TO RESOLVE

SUMMARY OF THE WORK DONE TODAY

IMPORTANT NOTES

NAME	SIGNATURE

TODAY LABOR

INITIALS	TRADE	START	FINISH	PAID HOURS	OVERTIME	COMPANY
☐ EMPLOYEE ☐ CONTRUCTOR		AM	PM			
☐ EMPLOYEE ☐ CONTRUCTOR		AM	PM			
☐ EMPLOYEE ☐ CONTRUCTOR		AM	PM			
☐ EMPLOYEE ☐ CONTRUCTOR		AM	PM			
☐ EMPLOYEE ☐ CONTRUCTOR		AM	PM			
☐ EMPLOYEE ☐ CONTRUCTOR		AM	PM			
☐ EMPLOYEE ☐ CONTRUCTOR		AM	PM			
☐ EMPLOYEE ☐ CONTRUCTOR		AM	PM			

EQUIPMENT ON SITE	NO. OF UNITE	WORKING YES / NO

HIRED EQUIPMENT	NO. OF UNITE	EQUIPMENT RENTED	FROM	RATE

NAME: _____   SIGNATURE: _____

MO	TU	WE	TH	FR	SA	SU		DATE:	/	/
☐	☐	☐	☐	☐	☐	☐				

PROJECT: **FOREMAN:**

WEATHER

F°_____ C°_____ _____ AM _____ PM

HOURS DUE TO BAD WEATHER	ISSUED AND DELAYS

NOTE: _____

COMPLETION DATE	DAYS AHEAD OF SCHEDULE	DAYS BEHIND SCHEDULE

SAFETY AND INCIDENTS

SAFETY ISSUES THAT NEED TO BE ADDRESSED	ACCIDENTS / INCIDENTS / STEPS NEEDED TO RESOLVE

SUMMARY OF THE WORK DONE TODAY

IMPORTANT NOTES

NAME	SIGNATURE

TODAY LABOR

INITIALS	TRADE	START	FINISH	PAID HOURS	OVERTIME	COMPANY
☐ EMPLOYEE ☐ CONTRUCTOR		AM	PM			
☐ EMPLOYEE ☐ CONTRUCTOR		AM	PM			
☐ EMPLOYEE ☐ CONTRUCTOR		AM	PM			
☐ EMPLOYEE ☐ CONTRUCTOR		AM	PM			
☐ EMPLOYEE ☐ CONTRUCTOR		AM	PM			
☐ EMPLOYEE ☐ CONTRUCTOR		AM	PM			
☐ EMPLOYEE ☐ CONTRUCTOR		AM	PM			
☐ EMPLOYEE ☐ CONTRUCTOR		AM	PM			

EQUIPMENT ON SITE	NO. OF UNITE	WORKING YES / NO

HIRED EQUIPMENT	NO. OF UNITE	EQUIPMENT RENTED	FROM	RATE

NAME: _____   SIGNATURE: _____

MO   TU   WE   TH   FR   SA   SU
☐   ☐   ☐   ☐   ☐   ☐   ☐   DATE: /   /

PROJECT: FOREMAN:

WEATHER   ☁🌧 ⛅ ☁ 🌨 ☀ 🌧 ⛈ | HOURS DUE TO | ISSUED AND DELAYS |
                                         | BAD WEATHER | |
           F°____ C°____ ____AM ____PM

NOTE: _____

COMPLETION DATE	DAYS AHEAD OF SCHEDULE	DAYS BEHIND SCHEDULE

SAFETY AND INCIDENTS

SAFETY ISSUES THAT NEED TO BE ADDRESSED	ACCIDENTS / INCIDENTS / STEPS NEEDED TO RESOLVE
_____	_____
_____	_____
_____	_____
_____	_____

SUMMARY OF THE WORK DONE TODAY

_____
_____
_____
_____
_____

IMPORTANT NOTES

_____
_____
_____
_____
_____

NAME	SIGNATURE

TODAY LABOR

INITIALS	TRADE	START	FINISH	PAID HOURS	OVERTIME	COMPANY
☐ EMPLOYEE ☐ CONTRUCTOR		AM	PM			
☐ EMPLOYEE ☐ CONTRUCTOR		AM	PM			
☐ EMPLOYEE ☐ CONTRUCTOR		AM	PM			
☐ EMPLOYEE ☐ CONTRUCTOR		AM	PM			
☐ EMPLOYEE ☐ CONTRUCTOR		AM	PM			
☐ EMPLOYEE ☐ CONTRUCTOR		AM	PM			
☐ EMPLOYEE ☐ CONTRUCTOR		AM	PM			
☐ EMPLOYEE ☐ CONTRUCTOR		AM	PM			

EQUIPMENT ON SITE	NO. OF UNITE	WORKING YES / NO

HIRED EQUIPMENT	NO. OF UNITE	EQUIPMENT RENTED	FROM	RATE

NAME: _____   SIGNATURE: _____

MO	TU	WE	TH	FR	SA	SU	DATE: / /
☐	☐	☐	☐	☐	☐	☐	

PROJECT: FOREMAN:

WEATHER

F°_____ C°_____   _____AM _____PM

HOURS DUE TO BAD WEATHER	ISSUED AND DELAYS

NOTE: _____

COMPLETION DATE	DAYS AHEAD OF SCHEDULE	DAYS BEHIND SCHEDULE

SAFETY AND INCIDENTS

SAFETY ISSUES THAT NEED TO BE ADDRESSED	ACCIDENTS / INCIDENTS / STEPS NEEDED TO RESOLVE
_____	_____
_____	_____
_____	_____
_____	_____

SUMMARY OF THE WORK DONE TODAY

_____
_____
_____
_____
_____
_____

IMPORTANT NOTES

_____
_____
_____
_____
_____

NAME	SIGNATURE

TODAY LABOR

INITIALS	TRADE	START	FINISH	PAID HOURS	OVERTIME	COMPANY
☐ EMPLOYEE ☐ CONTRUCTOR		AM	PM			
☐ EMPLOYEE ☐ CONTRUCTOR		AM	PM			
☐ EMPLOYEE ☐ CONTRUCTOR		AM	PM			
☐ EMPLOYEE ☐ CONTRUCTOR		AM	PM			
☐ EMPLOYEE ☐ CONTRUCTOR		AM	PM			
☐ EMPLOYEE ☐ CONTRUCTOR		AM	PM			
☐ EMPLOYEE ☐ CONTRUCTOR		AM	PM			
☐ EMPLOYEE ☐ CONTRUCTOR		AM	PM			

EQUIPMENT ON SITE	NO. OF UNITE	WORKING YES / NO

HIRED EQUIPMENT	NO. OF UNITE	EQUIPMENT RENTED	FROM	RATE

NAME: _____   SIGNATURE: _____

MO TU WE TH FR SA SU
☐   ☐   ☐   ☐   ☐   ☐   ☐

DATE: ____ / ____ / ____

PROJECT:

FOREMAN:

WEATHER ☁ ⛅ ☁ 🌠 ☀ 🌧 ⛈

F° ____ C° ____   ____ AM ____ PM

HOURS DUE TO BAD WEATHER	ISSUED AND DELAYS

NOTE: _____

COMPLETION DATE	DAYS AHEAD OF SCHEDULE	DAYS BEHIND SCHEDULE

SAFETY AND INCIDENTS

SAFETY ISSUES THAT NEED TO BE ADDRESSED	ACCIDENTS / INCIDENTS / STEPS NEEDED TO RESOLVE

SUMMARY OF THE WORK DONE TODAY

_____
_____
_____
_____
_____

IMPORTANT NOTES

_____
_____
_____
_____
_____

NAME	SIGNATURE

TODAY LABOR

INITIALS	TRADE	START	FINISH	PAID HOURS	OVERTIME	COMPANY
☐ EMPLOYEE ☐ CONTRUCTOR		AM	PM			
☐ EMPLOYEE ☐ CONTRUCTOR		AM	PM			
☐ EMPLOYEE ☐ CONTRUCTOR		AM	PM			
☐ EMPLOYEE ☐ CONTRUCTOR		AM	PM			
☐ EMPLOYEE ☐ CONTRUCTOR		AM	PM			
☐ EMPLOYEE ☐ CONTRUCTOR		AM	PM			
☐ EMPLOYEE ☐ CONTRUCTOR		AM	PM			
☐ EMPLOYEE ☐ CONTRUCTOR		AM	PM			

EQUIPMENT ON SITE	NO. OF UNITE	WORKING YES / NO

HIRED EQUIPMENT	NO. OF UNITE	EQUIPMENT RENTED	FROM	RATE

NAME: _____   SIGNATURE: _____

MO   TU   WE   TH   FR   SA   SU   DATE: /   /
☐   ☐   ☐   ☐   ☐   ☐   ☐

PROJECT: FOREMAN:

WEATHER

F°_____ C°_____   _____AM _____PM

| HOURS DUE TO BAD WEATHER | ISSUED AND DELAYS |

NOTE: _____

COMPLETION DATE	DAYS AHEAD OF SCHEDULE	DAYS BEHIND SCHEDULE

SAFETY AND INCIDENTS

SAFETY ISSUES THAT NEED TO BE ADDRESSED	ACCIDENTS / INCIDENTS / STEPS NEEDED TO RESOLVE

SUMMARY OF THE WORK DONE TODAY

IMPORTANT NOTES

NAME	SIGNATURE

TODAY LABOR

INITIALS	TRADE	START	FINISH	PAID HOURS	OVERTIME	COMPANY
☐ EMPLOYEE ☐ CONTRUCTOR		AM	PM			
☐ EMPLOYEE ☐ CONTRUCTOR		AM	PM			
☐ EMPLOYEE ☐ CONTRUCTOR		AM	PM			
☐ EMPLOYEE ☐ CONTRUCTOR		AM	PM			
☐ EMPLOYEE ☐ CONTRUCTOR		AM	PM			
☐ EMPLOYEE ☐ CONTRUCTOR		AM	PM			
☐ EMPLOYEE ☐ CONTRUCTOR		AM	PM			
☐ EMPLOYEE ☐ CONTRUCTOR		AM	PM			

EQUIPMENT ON SITE	NO. OF UNITE	WORKING YES / NO

HIRED EQUIPMENT	NO. OF UNITE	EQUIPMENT RENTED	FROM	RATE

NAME: _____   SIGNATURE: _____

MO TU WE TH FR SA SU
☐   ☐   ☐   ☐   ☐   ☐   ☐   DATE: /   /

PROJECT: _____   FOREMAN: _____

WEATHER [icons]

F°____   C°____   ____AM   ____PM

| HOURS DUE TO BAD WEATHER | ISSUED AND DELAYS |

NOTE: _____

COMPLETION DATE	DAYS AHEAD OF SCHEDULE	DAYS BEHIND SCHEDULE

SAFETY AND INCIDENTS

SAFETY ISSUES THAT NEED TO BE ADDRESSED	ACCIDENTS / INCIDENTS / STEPS NEEDED TO RESOLVE

SUMMARY OF THE WORK DONE TODAY

_____
_____
_____
_____
_____

IMPORTANT NOTES

_____
_____
_____
_____
_____

NAME	SIGNATURE

TODAY LABOR

INITIALS	TRADE	START	FINISH	PAID HOURS	OVERTIME	COMPANY
☐ EMPLOYEE ☐ CONTRUCTOR		AM	PM			
☐ EMPLOYEE ☐ CONTRUCTOR		AM	PM			
☐ EMPLOYEE ☐ CONTRUCTOR		AM	PM			
☐ EMPLOYEE ☐ CONTRUCTOR		AM	PM			
☐ EMPLOYEE ☐ CONTRUCTOR		AM	PM			
☐ EMPLOYEE ☐ CONTRUCTOR		AM	PM			
☐ EMPLOYEE ☐ CONTRUCTOR		AM	PM			
☐ EMPLOYEE ☐ CONTRUCTOR		AM	PM			

EQUIPMENT ON SITE	NO. OF UNITE	WORKING YES / NO

HIRED EQUIPMENT	NO. OF UNITE	EQUIPMENT RENTED	FROM	RATE

NAME: _____   SIGNATURE: _____

MO TU WE TH FR SA SU
☐   ☐   ☐   ☐   ☐   ☐   ☐

DATE: ___/___/___

PROJECT: _____

FOREMAN: _____

WEATHER

F°_____ C°_____   _____AM _____PM

HOURS DUE TO BAD WEATHER	ISSUED AND DELAYS

NOTE: _____

COMPLETION DATE	DAYS AHEAD OF SCHEDULE	DAYS BEHIND SCHEDULE

SAFETY AND INCIDENTS

SAFETY ISSUES THAT NEED TO BE ADDRESSED	ACCIDENTS / INCIDENTS / STEPS NEEDED TO RESOLVE

SUMMARY OF THE WORK DONE TODAY

_____
_____
_____
_____
_____

IMPORTANT NOTES

_____
_____
_____
_____
_____

NAME	SIGNATURE

TODAY LABOR

INITIALS	TRADE	START	FINISH	PAID HOURS	OVERTIME	COMPANY
☐ EMPLOYEE ☐ CONTRUCTOR		AM	PM			
☐ EMPLOYEE ☐ CONTRUCTOR		AM	PM			
☐ EMPLOYEE ☐ CONTRUCTOR		AM	PM			
☐ EMPLOYEE ☐ CONTRUCTOR		AM	PM			
☐ EMPLOYEE ☐ CONTRUCTOR		AM	PM			
☐ EMPLOYEE ☐ CONTRUCTOR		AM	PM			
☐ EMPLOYEE ☐ CONTRUCTOR		AM	PM			
☐ EMPLOYEE ☐ CONTRUCTOR		AM	PM			

EQUIPMENT ON SITE	NO. OF UNITE	WORKING YES / NO

HIRED EQUIPMENT	NO. OF UNITE	EQUIPMENT RENTED	FROM	RATE

NAME: _____   SIGNATURE: _____

MO   TU   WE   TH   FR   SA   SU
☐   ☐   ☐   ☐   ☐   ☐   ☐

DATE: /   /

PROJECT:

FOREMAN:

WEATHER

F°_____ C°_____   _____ AM   _____ PM

HOURS DUE TO BAD WEATHER	ISSUED AND DELAYS

NOTE: _____

COMPLETION DATE	DAYS AHEAD OF SCHEDULE	DAYS BEHIND SCHEDULE

SAFETY AND INCIDENTS

SAFETY ISSUES THAT NEED TO BE ADDRESSED	ACCIDENTS / INCIDENTS / STEPS NEEDED TO RESOLVE

SUMMARY OF THE WORK DONE TODAY

IMPORTANT NOTES

NAME	SIGNATURE

TODAY LABOR

INITIALS	TRADE	START	FINISH	PAID HOURS	OVERTIME	COMPANY
☐ EMPLOYEE ☐ CONTRUCTOR		AM	PM			
☐ EMPLOYEE ☐ CONTRUCTOR		AM	PM			
☐ EMPLOYEE ☐ CONTRUCTOR		AM	PM			
☐ EMPLOYEE ☐ CONTRUCTOR		AM	PM			
☐ EMPLOYEE ☐ CONTRUCTOR		AM	PM			
☐ EMPLOYEE ☐ CONTRUCTOR		AM	PM			
☐ EMPLOYEE ☐ CONTRUCTOR		AM	PM			
☐ EMPLOYEE ☐ CONTRUCTOR		AM	PM			

EQUIPMENT ON SITE	NO. OF UNITE	WORKING YES / NO

HIRED EQUIPMENT	NO. OF UNITE	EQUIPMENT RENTED	FROM	RATE

NAME: _____   SIGNATURE: _____

MO   TU   WE   TH   FR   SA   SU
☐   ☐   ☐   ☐   ☐   ☐   ☐

DATE: /   /

PROJECT:

FOREMAN:

WEATHER

F°_____ C°_____   _____ AM   _____ PM

HOURS DUE TO BAD WEATHER	ISSUED AND DELAYS

NOTE: _____

COMPLETION DATE	DAYS AHEAD OF SCHEDULE	DAYS BEHIND SCHEDULE

SAFETY AND INCIDENTS

SAFETY ISSUES THAT NEED TO BE ADDRESSED	ACCIDENTS / INCIDENTS / STEPS NEEDED TO RESOLVE

SUMMARY OF THE WORK DONE TODAY

IMPORTANT NOTES

NAME	SIGNATURE

TODAY LABOR

INITIALS	TRADE	START	FINISH	PAID HOURS	OVERTIME	COMPANY
☐ EMPLOYEE ☐ CONTRUCTOR		AM	PM			
☐ EMPLOYEE ☐ CONTRUCTOR		AM	PM			
☐ EMPLOYEE ☐ CONTRUCTOR		AM	PM			
☐ EMPLOYEE ☐ CONTRUCTOR		AM	PM			
☐ EMPLOYEE ☐ CONTRUCTOR		AM	PM			
☐ EMPLOYEE ☐ CONTRUCTOR		AM	PM			
☐ EMPLOYEE ☐ CONTRUCTOR		AM	PM			
☐ EMPLOYEE ☐ CONTRUCTOR		AM	PM			

EQUIPMENT ON SITE	NO. OF UNITE	WORKING YES / NO

HIRED EQUIPMENT	NO. OF UNITE	EQUIPMENT RENTED	FROM	RATE

NAME: _____   SIGNATURE: _____

MO TU WE TH FR SA SU
☐   ☐   ☐   ☐   ☐   ☐   ☐

DATE: /   /

PROJECT:

FOREMAN:

WEATHER

F°_____ C°_____   _____ AM _____ PM

HOURS DUE TO
BAD WEATHER

ISSUED AND DELAYS

NOTE: _____

COMPLETION DATE	DAYS AHEAD OF SCHEDULE	DAYS BEHIND SCHEDULE

SAFETY AND INCIDENTS

SAFETY ISSUES THAT NEED TO BE ADDRESSED	ACCIDENTS / INCIDENTS / STEPS NEEDED TO RESOLVE

SUMMARY OF THE WORK DONE TODAY

IMPORTANT NOTES

NAME	SIGNATURE

TODAY LABOR

INITIALS	TRADE	START	FINISH	PAID HOURS	OVERTIME	COMPANY
☐ EMPLOYEE ☐ CONTRUCTOR		AM	PM			
☐ EMPLOYEE ☐ CONTRUCTOR		AM	PM			
☐ EMPLOYEE ☐ CONTRUCTOR		AM	PM			
☐ EMPLOYEE ☐ CONTRUCTOR		AM	PM			
☐ EMPLOYEE ☐ CONTRUCTOR		AM	PM			
☐ EMPLOYEE ☐ CONTRUCTOR		AM	PM			
☐ EMPLOYEE ☐ CONTRUCTOR		AM	PM			
☐ EMPLOYEE ☐ CONTRUCTOR		AM	PM			

EQUIPMENT ON SITE	NO. OF UNITE	WORKING YES / NO

HIRED EQUIPMENT	NO. OF UNITE	EQUIPMENT RENTED	FROM	RATE

NAME: _____   SIGNATURE: _____

MO TU WE TH FR SA SU
☐   ☐   ☐   ☐   ☐   ☐   ☐

DATE: ___/___/___

PROJECT: _____

FOREMAN: _____

WEATHER

F°____ C°____ ____AM ____PM

HOURS DUE TO BAD WEATHER	ISSUED AND DELAYS

NOTE: _____

COMPLETION DATE	DAYS AHEAD OF SCHEDULE	DAYS BEHIND SCHEDULE

SAFETY AND INCIDENTS

SAFETY ISSUES THAT NEED TO BE ADDRESSED	ACCIDENTS / INCIDENTS / STEPS NEEDED TO RESOLVE

SUMMARY OF THE WORK DONE TODAY

IMPORTANT NOTES

NAME	SIGNATURE

TODAY LABOR

INITIALS	TRADE	START	FINISH	PAID HOURS	OVERTIME	COMPANY
☐ EMPLOYEE ☐ CONTRUCTOR		AM	PM			
☐ EMPLOYEE ☐ CONTRUCTOR		AM	PM			
☐ EMPLOYEE ☐ CONTRUCTOR		AM	PM			
☐ EMPLOYEE ☐ CONTRUCTOR		AM	PM			
☐ EMPLOYEE ☐ CONTRUCTOR		AM	PM			
☐ EMPLOYEE ☐ CONTRUCTOR		AM	PM			
☐ EMPLOYEE ☐ CONTRUCTOR		AM	PM			
☐ EMPLOYEE ☐ CONTRUCTOR		AM	PM			

EQUIPMENT ON SITE	NO. OF UNITE	WORKING YES / NO

HIRED EQUIPMENT	NO. OF UNITE	EQUIPMENT RENTED	FROM	RATE

NAME: _____   SIGNATURE: _____

MO TU WE TH FR SA SU
☐   ☐   ☐   ☐   ☐   ☐   ☐

DATE: ___ / ___ / ___

PROJECT: _____

FOREMAN: _____

WEATHER ☁☂ ⛅ ☁ 🌨 ☀ 🌧 ⛈

F°_____ C°_____ _____AM _____PM

HOURS DUE TO BAD WEATHER	ISSUED AND DELAYS

NOTE: _____

COMPLETION DATE	DAYS AHEAD OF SCHEDULE	DAYS BEHIND SCHEDULE

SAFETY AND INCIDENTS

SAFETY ISSUES THAT NEED TO BE ADDRESSED	ACCIDENTS / INCIDENTS / STEPS NEEDED TO RESOLVE

SUMMARY OF THE WORK DONE TODAY

_____
_____
_____
_____
_____

IMPORTANT NOTES

_____
_____
_____
_____
_____

NAME	SIGNATURE

TODAY LABOR

INITIALS	TRADE	START	FINISH	PAID HOURS	OVERTIME	COMPANY
☐ EMPLOYEE ☐ CONTRUCTOR		AM	PM			
☐ EMPLOYEE ☐ CONTRUCTOR		AM	PM			
☐ EMPLOYEE ☐ CONTRUCTOR		AM	PM			
☐ EMPLOYEE ☐ CONTRUCTOR		AM	PM			
☐ EMPLOYEE ☐ CONTRUCTOR		AM	PM			
☐ EMPLOYEE ☐ CONTRUCTOR		AM	PM			
☐ EMPLOYEE ☐ CONTRUCTOR		AM	PM			
☐ EMPLOYEE ☐ CONTRUCTOR		AM	PM			

EQUIPMENT ON SITE	NO. OF UNITE	WORKING YES / NO

HIRED EQUIPMENT	NO. OF UNITE	EQUIPMENT RENTED	FROM	RATE

NAME: _____   SIGNATURE: _____

MO TU WE TH FR SA SU
☐   ☐   ☐   ☐   ☐   ☐   ☐

DATE: /   /

PROJECT:

FOREMAN:

WEATHER

F°____ C°____   ____ AM ____ PM

HOURS DUE TO
BAD WEATHER

ISSUED AND DELAYS

NOTE: _____

COMPLETION DATE	DAYS AHEAD OF SCHEDULE	DAYS BEHIND SCHEDULE

SAFETY AND INCIDENTS

SAFETY ISSUES THAT NEED TO BE ADDRESSED	ACCIDENTS / INCIDENTS / STEPS NEEDED TO RESOLVE

SUMMARY OF THE WORK DONE TODAY

IMPORTANT NOTES

NAME	SIGNATURE

TODAY LABOR

INITIALS	TRADE	START	FINISH	PAID HOURS	OVERTIME	COMPANY
☐ EMPLOYEE ☐ CONTRUCTOR		AM	PM			
☐ EMPLOYEE ☐ CONTRUCTOR		AM	PM			
☐ EMPLOYEE ☐ CONTRUCTOR		AM	PM			
☐ EMPLOYEE ☐ CONTRUCTOR		AM	PM			
☐ EMPLOYEE ☐ CONTRUCTOR		AM	PM			
☐ EMPLOYEE ☐ CONTRUCTOR		AM	PM			
☐ EMPLOYEE ☐ CONTRUCTOR		AM	PM			
☐ EMPLOYEE ☐ CONTRUCTOR		AM	PM			

EQUIPMENT ON SITE	NO. OF UNITE	WORKING YES / NO

HIRED EQUIPMENT	NO. OF UNITE	EQUIPMENT RENTED	FROM	RATE

NAME: _____   SIGNATURE: _____

MO TU WE TH FR SA SU
☐   ☐   ☐   ☐   ☐   ☐   ☐

DATE: __ / __ / __

PROJECT:

FOREMAN:

WEATHER   F°____ C°____ ____AM ____PM

HOURS DUE TO BAD WEATHER	ISSUED AND DELAYS

NOTE: _____

COMPLETION DATE	DAYS AHEAD OF SCHEDULE	DAYS BEHIND SCHEDULE

SAFETY AND INCIDENTS

SAFETY ISSUES THAT NEED TO BE ADDRESSED	ACCIDENTS / INCIDENTS / STEPS NEEDED TO RESOLVE

SUMMARY OF THE WORK DONE TODAY

IMPORTANT NOTES

NAME	SIGNATURE

TODAY LABOR

INITIALS	TRADE	START	FINISH	PAID HOURS	OVERTIME	COMPANY
☐ EMPLOYEE ☐ CONTRUCTOR		AM	PM			
☐ EMPLOYEE ☐ CONTRUCTOR		AM	PM			
☐ EMPLOYEE ☐ CONTRUCTOR		AM	PM			
☐ EMPLOYEE ☐ CONTRUCTOR		AM	PM			
☐ EMPLOYEE ☐ CONTRUCTOR		AM	PM			
☐ EMPLOYEE ☐ CONTRUCTOR		AM	PM			
☐ EMPLOYEE ☐ CONTRUCTOR		AM	PM			
☐ EMPLOYEE ☐ CONTRUCTOR		AM	PM			

EQUIPMENT ON SITE	NO. OF UNITE	WORKING YES / NO

HIRED EQUIPMENT	NO. OF UNITE	EQUIPMENT RENTED	FROM	RATE

NAME: _____   SIGNATURE: _____

MO TU WE TH FR SA SU
☐   ☐   ☐   ☐   ☐   ☐   ☐   DATE: /   /

PROJECT: FOREMAN:

WEATHER   ☁ ⛅ ☁ 🌧 ☀ 🌧 ⛈ | HOURS DUE TO | ISSUED AND DELAYS |
                                        | BAD WEATHER | |
           F°___ C°___   ___ AM ___ PM

NOTE: _____

COMPLETION DATE	DAYS AHEAD OF SCHEDULE	DAYS BEHIND SCHEDULE

SAFETY AND INCIDENTS

SAFETY ISSUES THAT NEED TO BE ADDRESSED	ACCIDENTS / INCIDENTS / STEPS NEEDED TO RESOLVE

SUMMARY OF THE WORK DONE TODAY

_____
_____
_____
_____
_____

IMPORTANT NOTES

_____
_____
_____
_____
_____

NAME	SIGNATURE

TODAY LABOR

INITIALS	TRADE	START	FINISH	PAID HOURS	OVERTIME	COMPANY
☐ EMPLOYEE ☐ CONTRUCTOR		AM	PM			
☐ EMPLOYEE ☐ CONTRUCTOR		AM	PM			
☐ EMPLOYEE ☐ CONTRUCTOR		AM	PM			
☐ EMPLOYEE ☐ CONTRUCTOR		AM	PM			
☐ EMPLOYEE ☐ CONTRUCTOR		AM	PM			
☐ EMPLOYEE ☐ CONTRUCTOR		AM	PM			
☐ EMPLOYEE ☐ CONTRUCTOR		AM	PM			
☐ EMPLOYEE ☐ CONTRUCTOR		AM	PM			

EQUIPMENT ON SITE	NO. OF UNITE	WORKING YES / NO

HIRED EQUIPMENT	NO. OF UNITE	EQUIPMENT RENTED	FROM	RATE

NAME: _____   SIGNATURE: _____

MO TU WE TH FR SA SU
☐   ☐   ☐   ☐   ☐   ☐   ☐

DATE: ___ / ___ / ___

PROJECT: _____

FOREMAN: _____

WEATHER ⛈ 🌤 ☁ 🌨 ☀ 🌧 ⛈

F° _____ C° _____ _____ AM _____ PM

HOURS DUE TO BAD WEATHER	ISSUED AND DELAYS

NOTE: _____

COMPLETION DATE	DAYS AHEAD OF SCHEDULE	DAYS BEHIND SCHEDULE

SAFETY AND INCIDENTS

SAFETY ISSUES THAT NEED TO BE ADDRESSED	ACCIDENTS / INCIDENTS / STEPS NEEDED TO RESOLVE

SUMMARY OF THE WORK DONE TODAY

IMPORTANT NOTES

NAME	SIGNATURE

TODAY LABOR

INITIALS	TRADE	START	FINISH	PAID HOURS	OVERTIME	COMPANY
☐ EMPLOYEE ☐ CONTRUCTOR		AM	PM			
☐ EMPLOYEE ☐ CONTRUCTOR		AM	PM			
☐ EMPLOYEE ☐ CONTRUCTOR		AM	PM			
☐ EMPLOYEE ☐ CONTRUCTOR		AM	PM			
☐ EMPLOYEE ☐ CONTRUCTOR		AM	PM			
☐ EMPLOYEE ☐ CONTRUCTOR		AM	PM			
☐ EMPLOYEE ☐ CONTRUCTOR		AM	PM			
☐ EMPLOYEE ☐ CONTRUCTOR		AM	PM			

EQUIPMENT ON SITE	NO. OF UNITE	WORKING YES / NO

HIRED EQUIPMENT	NO. OF UNITE	EQUIPMENT RENTED	FROM	RATE

NAME: _____   SIGNATURE: _____

| MO | TU | WE | TH | FR | SA | SU |
| ☐ | ☐ | ☐ | ☐ | ☐ | ☐ | ☐ |

DATE: / /

PROJECT:

FOREMAN:

WEATHER

F°_____ C°_____ _____ AM _____ PM

| HOURS DUE TO BAD WEATHER | ISSUED AND DELAYS |

NOTE: _____

| COMPLETION DATE | DAYS AHEAD OF SCHEDULE | DAYS BEHIND SCHEDULE |

SAFETY AND INCIDENTS

| SAFETY ISSUES THAT NEED TO BE ADDRESSED | ACCIDENTS / INCIDENTS / STEPS NEEDED TO RESOLVE |

SUMMARY OF THE WORK DONE TODAY

IMPORTANT NOTES

| NAME | SIGNATURE |

TODAY LABOR

INITIALS	TRADE	START	FINISH	PAID HOURS	OVERTIME	COMPANY
☐ EMPLOYEE ☐ CONTRUCTOR		AM	PM			
☐ EMPLOYEE ☐ CONTRUCTOR		AM	PM			
☐ EMPLOYEE ☐ CONTRUCTOR		AM	PM			
☐ EMPLOYEE ☐ CONTRUCTOR		AM	PM			
☐ EMPLOYEE ☐ CONTRUCTOR		AM	PM			
☐ EMPLOYEE ☐ CONTRUCTOR		AM	PM			
☐ EMPLOYEE ☐ CONTRUCTOR		AM	PM			
☐ EMPLOYEE ☐ CONTRUCTOR		AM	PM			

EQUIPMENT ON SITE	NO. OF UNITE	WORKING YES / NO

HIRED EQUIPMENT	NO. OF UNITE	EQUIPMENT RENTED	FROM	RATE

NAME: _____   SIGNATURE: _____

MO   TU   WE   TH   FR   SA   SU
☐   ☐   ☐   ☐   ☐   ☐   ☐

DATE: ___ / ___ / ___

PROJECT:

FOREMAN:

WEATHER

F°_____ C°_____   _____AM _____PM

HOURS DUE TO BAD WEATHER	ISSUED AND DELAYS

NOTE: _____

COMPLETION DATE	DAYS AHEAD OF SCHEDULE	DAYS BEHIND SCHEDULE

SAFETY AND INCIDENTS

SAFETY ISSUES THAT NEED TO BE ADDRESSED	ACCIDENTS / INCIDENTS / STEPS NEEDED TO RESOLVE

SUMMARY OF THE WORK DONE TODAY

IMPORTANT NOTES

NAME	SIGNATURE

TODAY LABOR

INITIALS	TRADE	START	FINISH	PAID HOURS	OVERTIME	COMPANY
☐ EMPLOYEE ☐ CONTRUCTOR		AM	PM			
☐ EMPLOYEE ☐ CONTRUCTOR		AM	PM			
☐ EMPLOYEE ☐ CONTRUCTOR		AM	PM			
☐ EMPLOYEE ☐ CONTRUCTOR		AM	PM			
☐ EMPLOYEE ☐ CONTRUCTOR		AM	PM			
☐ EMPLOYEE ☐ CONTRUCTOR		AM	PM			
☐ EMPLOYEE ☐ CONTRUCTOR		AM	PM			
☐ EMPLOYEE ☐ CONTRUCTOR		AM	PM			

EQUIPMENT ON SITE	NO. OF UNITE	WORKING YES / NO

HIRED EQUIPMENT	NO. OF UNITE	EQUIPMENT RENTED	FROM	RATE

NAME: _____   SIGNATURE: _____

MO	TU	WE	TH	FR	SA	SU		DATE:	/	/
☐	☐	☐	☐	☐	☐	☐				

PROJECT:

FOREMAN:

WEATHER

F°_____ C°_____ _____AM _____PM

HOURS DUE TO BAD WEATHER	ISSUED AND DELAYS

NOTE: _____

COMPLETION DATE	DAYS AHEAD OF SCHEDULE	DAYS BEHIND SCHEDULE

SAFETY AND INCIDENTS

SAFETY ISSUES THAT NEED TO BE ADDRESSED	ACCIDENTS / INCIDENTS / STEPS NEEDED TO RESOLVE

SUMMARY OF THE WORK DONE TODAY

IMPORTANT NOTES

NAME	SIGNATURE

TODAY LABOR

INITIALS	TRADE	START	FINISH	PAID HOURS	OVERTIME	COMPANY
☐ EMPLOYEE ☐ CONTRUCTOR		AM	PM			
☐ EMPLOYEE ☐ CONTRUCTOR		AM	PM			
☐ EMPLOYEE ☐ CONTRUCTOR		AM	PM			
☐ EMPLOYEE ☐ CONTRUCTOR		AM	PM			
☐ EMPLOYEE ☐ CONTRUCTOR		AM	PM			
☐ EMPLOYEE ☐ CONTRUCTOR		AM	PM			
☐ EMPLOYEE ☐ CONTRUCTOR		AM	PM			
☐ EMPLOYEE ☐ CONTRUCTOR		AM	PM			

EQUIPMENT ON SITE	NO. OF UNITE	WORKING YES / NO

HIRED EQUIPMENT	NO. OF UNITE	EQUIPMENT RENTED	FROM	RATE

NAME: _____   SIGNATURE: _____

MO TU WE TH FR SA SU
☐   ☐   ☐   ☐   ☐   ☐   ☐

DATE: /   /

PROJECT:

FOREMAN:

WEATHER

F°_____ C°_____   _____AM _____PM

HOURS DUE TO BAD WEATHER	ISSUED AND DELAYS

NOTE: _____

COMPLETION DATE	DAYS AHEAD OF SCHEDULE	DAYS BEHIND SCHEDULE

SAFETY AND INCIDENTS

SAFETY ISSUES THAT NEED TO BE ADDRESSED	ACCIDENTS / INCIDENTS / STEPS NEEDED TO RESOLVE

SUMMARY OF THE WORK DONE TODAY

_____
_____
_____
_____
_____

IMPORTANT NOTES

_____
_____
_____
_____
_____

NAME	SIGNATURE

TODAY LABOR

INITIALS	TRADE	START	FINISH	PAID HOURS	OVERTIME	COMPANY
☐ EMPLOYEE ☐ CONTRUCTOR		AM	PM			
☐ EMPLOYEE ☐ CONTRUCTOR		AM	PM			
☐ EMPLOYEE ☐ CONTRUCTOR		AM	PM			
☐ EMPLOYEE ☐ CONTRUCTOR		AM	PM			
☐ EMPLOYEE ☐ CONTRUCTOR		AM	PM			
☐ EMPLOYEE ☐ CONTRUCTOR		AM	PM			
☐ EMPLOYEE ☐ CONTRUCTOR		AM	PM			
☐ EMPLOYEE ☐ CONTRUCTOR		AM	PM			

EQUIPMENT ON SITE	NO. OF UNITE	WORKING YES / NO

HIRED EQUIPMENT	NO. OF UNITE	EQUIPMENT RENTED	FROM	RATE

NAME: _____   SIGNATURE: _____

MO TU WE TH FR SA SU
☐   ☐   ☐   ☐   ☐   ☐   ☐

DATE: /   /

PROJECT:

FOREMAN:

WEATHER

F°_____ C°_____   _____ AM _____ PM

| HOURS DUE TO BAD WEATHER | ISSUED AND DELAYS |

NOTE: _____

COMPLETION DATE	DAYS AHEAD OF SCHEDULE	DAYS BEHIND SCHEDULE

SAFETY AND INCIDENTS

SAFETY ISSUES THAT NEED TO BE ADDRESSED	ACCIDENTS / INCIDENTS / STEPS NEEDED TO RESOLVE

SUMMARY OF THE WORK DONE TODAY

IMPORTANT NOTES

NAME	SIGNATURE

TODAY LABOR

INITIALS	TRADE	START	FINISH	PAID HOURS	OVERTIME	COMPANY
☐ EMPLOYEE ☐ CONTRUCTOR		AM	PM			
☐ EMPLOYEE ☐ CONTRUCTOR		AM	PM			
☐ EMPLOYEE ☐ CONTRUCTOR		AM	PM			
☐ EMPLOYEE ☐ CONTRUCTOR		AM	PM			
☐ EMPLOYEE ☐ CONTRUCTOR		AM	PM			
☐ EMPLOYEE ☐ CONTRUCTOR		AM	PM			
☐ EMPLOYEE ☐ CONTRUCTOR		AM	PM			
☐ EMPLOYEE ☐ CONTRUCTOR		AM	PM			

EQUIPMENT ON SITE	NO. OF UNITE	WORKING YES / NO

HIRED EQUIPMENT	NO. OF UNITE	EQUIPMENT RENTED	FROM	RATE

NAME: _____ SIGNATURE: _____

MO TU WE TH FR SA SU
☐   ☐   ☐   ☐   ☐   ☐   ☐

DATE: ___ / ___ / ___

PROJECT: _____

FOREMAN: _____

WEATHER

F°____ C°____ ____ AM ____ PM

HOURS DUE TO BAD WEATHER	ISSUED AND DELAYS

NOTE: _____

COMPLETION DATE	DAYS AHEAD OF SCHEDULE	DAYS BEHIND SCHEDULE

SAFETY AND INCIDENTS

SAFETY ISSUES THAT NEED TO BE ADDRESSED	ACCIDENTS / INCIDENTS / STEPS NEEDED TO RESOLVE

SUMMARY OF THE WORK DONE TODAY

IMPORTANT NOTES

NAME	SIGNATURE

TODAY LABOR

INITIALS	TRADE	START	FINISH	PAID HOURS	OVERTIME	COMPANY
☐ EMPLOYEE ☐ CONTRUCTOR		AM	PM			
☐ EMPLOYEE ☐ CONTRUCTOR		AM	PM			
☐ EMPLOYEE ☐ CONTRUCTOR		AM	PM			
☐ EMPLOYEE ☐ CONTRUCTOR		AM	PM			
☐ EMPLOYEE ☐ CONTRUCTOR		AM	PM			
☐ EMPLOYEE ☐ CONTRUCTOR		AM	PM			
☐ EMPLOYEE ☐ CONTRUCTOR		AM	PM			
☐ EMPLOYEE ☐ CONTRUCTOR		AM	PM			

EQUIPMENT ON SITE	NO. OF UNITE	WORKING YES / NO

HIRED EQUIPMENT	NO. OF UNITE	EQUIPMENT RENTED	FROM	RATE

NAME: _____   SIGNATURE: _____

MO TU WE TH FR SA SU
☐   ☐   ☐   ☐   ☐   ☐   ☐

DATE: /   /

PROJECT: FOREMAN:

WEATHER

F°_____ C°_____   _____ AM _____ PM

HOURS DUE TO BAD WEATHER	ISSUED AND DELAYS

NOTE: _____

COMPLETION DATE	DAYS AHEAD OF SCHEDULE	DAYS BEHIND SCHEDULE

SAFETY AND INCIDENTS

SAFETY ISSUES THAT NEED TO BE ADDRESSED	ACCIDENTS / INCIDENTS / STEPS NEEDED TO RESOLVE

SUMMARY OF THE WORK DONE TODAY

_____
_____
_____
_____
_____
_____

IMPORTANT NOTES

_____
_____
_____
_____
_____
_____

NAME	SIGNATURE

TODAY LABOR

INITIALS	TRADE	START	FINISH	PAID HOURS	OVERTIME	COMPANY
☐ EMPLOYEE ☐ CONTRUCTOR		AM	PM			
☐ EMPLOYEE ☐ CONTRUCTOR		AM	PM			
☐ EMPLOYEE ☐ CONTRUCTOR		AM	PM			
☐ EMPLOYEE ☐ CONTRUCTOR		AM	PM			
☐ EMPLOYEE ☐ CONTRUCTOR		AM	PM			
☐ EMPLOYEE ☐ CONTRUCTOR		AM	PM			
☐ EMPLOYEE ☐ CONTRUCTOR		AM	PM			
☐ EMPLOYEE ☐ CONTRUCTOR		AM	PM			

EQUIPMENT ON SITE	NO. OF UNITE	WORKING YES / NO

HIRED EQUIPMENT	NO. OF UNITE	EQUIPMENT RENTED	FROM	RATE

NAME: _____   SIGNATURE: _____

MO   TU   WE   TH   FR   SA   SU

☐   ☐   ☐   ☐   ☐   ☐   ☐

DATE: ___/___/___

PROJECT: _____

FOREMAN: _____

WEATHER   ☁   ⛅   ☁   🌨   ☀   🌧   ⛈

F°_____ C°_____   _____AM   _____PM

HOURS DUE TO BAD WEATHER	ISSUED AND DELAYS

NOTE: _____

COMPLETION DATE	DAYS AHEAD OF SCHEDULE	DAYS BEHIND SCHEDULE

SAFETY AND INCIDENTS

SAFETY ISSUES THAT NEED TO BE ADDRESSED	ACCIDENTS / INCIDENTS / STEPS NEEDED TO RESOLVE

SUMMARY OF THE WORK DONE TODAY

_____
_____
_____
_____
_____

IMPORTANT NOTES

_____
_____
_____
_____
_____

NAME	SIGNATURE

TODAY LABOR

INITIALS	TRADE	START	FINISH	PAID HOURS	OVERTIME	COMPANY
☐ EMPLOYEE ☐ CONTRUCTOR		AM	PM			
☐ EMPLOYEE ☐ CONTRUCTOR		AM	PM			
☐ EMPLOYEE ☐ CONTRUCTOR		AM	PM			
☐ EMPLOYEE ☐ CONTRUCTOR		AM	PM			
☐ EMPLOYEE ☐ CONTRUCTOR		AM	PM			
☐ EMPLOYEE ☐ CONTRUCTOR		AM	PM			
☐ EMPLOYEE ☐ CONTRUCTOR		AM	PM			
☐ EMPLOYEE ☐ CONTRUCTOR		AM	PM			

EQUIPMENT ON SITE	NO. OF UNITE	WORKING YES / NO

HIRED EQUIPMENT	NO. OF UNITE	EQUIPMENT RENTED	FROM	RATE

NAME: _____   SIGNATURE: _____

MO TU WE TH FR SA SU
☐   ☐   ☐   ☐   ☐   ☐   ☐

DATE: ___ / ___ / ___

PROJECT:

FOREMAN:

WEATHER   F° ____   C° ____   ____ AM   ____ PM

HOURS DUE TO BAD WEATHER	ISSUED AND DELAYS

NOTE:

COMPLETION DATE	DAYS AHEAD OF SCHEDULE	DAYS BEHIND SCHEDULE

SAFETY AND INCIDENTS

SAFETY ISSUES THAT NEED TO BE ADDRESSED	ACCIDENTS / INCIDENTS / STEPS NEEDED TO RESOLVE

SUMMARY OF THE WORK DONE TODAY

IMPORTANT NOTES

NAME	SIGNATURE

TODAY LABOR

INITIALS	TRADE	START	FINISH	PAID HOURS	OVERTIME	COMPANY
☐ EMPLOYEE ☐ CONTRUCTOR		AM	PM			
☐ EMPLOYEE ☐ CONTRUCTOR		AM	PM			
☐ EMPLOYEE ☐ CONTRUCTOR		AM	PM			
☐ EMPLOYEE ☐ CONTRUCTOR		AM	PM			
☐ EMPLOYEE ☐ CONTRUCTOR		AM	PM			
☐ EMPLOYEE ☐ CONTRUCTOR		AM	PM			
☐ EMPLOYEE ☐ CONTRUCTOR		AM	PM			
☐ EMPLOYEE ☐ CONTRUCTOR		AM	PM			

EQUIPMENT ON SITE	NO. OF UNITE	WORKING YES / NO

HIRED EQUIPMENT	NO. OF UNITE	EQUIPMENT RENTED	FROM	RATE

NAME: _____   SIGNATURE: _____

MO	TU	WE	TH	FR	SA	SU	DATE: / /
☐	☐	☐	☐	☐	☐	☐	

PROJECT:

FOREMAN:

WEATHER ☁ ⛅ ☁ 🌦 ☀ 🌧 ⛈

F°_____ C°_____ _____AM _____PM

HOURS DUE TO BAD WEATHER	ISSUED AND DELAYS

NOTE: _____

COMPLETION DATE	DAYS AHEAD OF SCHEDULE	DAYS BEHIND SCHEDULE

SAFETY AND INCIDENTS

SAFETY ISSUES THAT NEED TO BE ADDRESSED	ACCIDENTS / INCIDENTS / STEPS NEEDED TO RESOLVE

SUMMARY OF THE WORK DONE TODAY

IMPORTANT NOTES

NAME	SIGNATURE

TODAY LABOR

INITIALS	TRADE	START	FINISH	PAID HOURS	OVERTIME	COMPANY
☐ EMPLOYEE ☐ CONTRUCTOR		AM	PM			
☐ EMPLOYEE ☐ CONTRUCTOR		AM	PM			
☐ EMPLOYEE ☐ CONTRUCTOR		AM	PM			
☐ EMPLOYEE ☐ CONTRUCTOR		AM	PM			
☐ EMPLOYEE ☐ CONTRUCTOR		AM	PM			
☐ EMPLOYEE ☐ CONTRUCTOR		AM	PM			
☐ EMPLOYEE ☐ CONTRUCTOR		AM	PM			
☐ EMPLOYEE ☐ CONTRUCTOR		AM	PM			

EQUIPMENT ON SITE	NO. OF UNITE	WORKING YES / NO

HIRED EQUIPMENT	NO. OF UNITE	EQUIPMENT RENTED	FROM	RATE

NAME: _____   SIGNATURE: _____

MO TU WE TH FR SA SU
☐   ☐   ☐   ☐   ☐   ☐   ☐

DATE: /   /

PROJECT:

FOREMAN:

WEATHER

F°_____ C°_____   _____ AM _____ PM

| HOURS DUE TO BAD WEATHER | ISSUED AND DELAYS |

NOTE: _____

COMPLETION DATE	DAYS AHEAD OF SCHEDULE	DAYS BEHIND SCHEDULE

SAFETY AND INCIDENTS

SAFETY ISSUES THAT NEED TO BE ADDRESSED	ACCIDENTS / INCIDENTS / STEPS NEEDED TO RESOLVE

SUMMARY OF THE WORK DONE TODAY

_____
_____
_____
_____
_____
_____

IMPORTANT NOTES

_____
_____
_____
_____
_____
_____

NAME	SIGNATURE

TODAY LABOR

INITIALS	TRADE	START	FINISH	PAID HOURS	OVERTIME	COMPANY
☐ EMPLOYEE ☐ CONTRUCTOR		AM	PM			
☐ EMPLOYEE ☐ CONTRUCTOR		AM	PM			
☐ EMPLOYEE ☐ CONTRUCTOR		AM	PM			
☐ EMPLOYEE ☐ CONTRUCTOR		AM	PM			
☐ EMPLOYEE ☐ CONTRUCTOR		AM	PM			
☐ EMPLOYEE ☐ CONTRUCTOR		AM	PM			
☐ EMPLOYEE ☐ CONTRUCTOR		AM	PM			
☐ EMPLOYEE ☐ CONTRUCTOR		AM	PM			

EQUIPMENT ON SITE	NO. OF UNITE	WORKING YES / NO

HIRED EQUIPMENT	NO. OF UNITE	EQUIPMENT RENTED	FROM	RATE

NAME: _____   SIGNATURE: _____

MO TU WE TH FR SA SU
☐   ☐   ☐   ☐   ☐   ☐   ☐

DATE: ___ / ___ / ___

PROJECT:

FOREMAN:

WEATHER

F°_____ C°_____   _____ AM _____ PM

HOURS DUE TO BAD WEATHER	ISSUED AND DELAYS

NOTE: _____

COMPLETION DATE	DAYS AHEAD OF SCHEDULE	DAYS BEHIND SCHEDULE

SAFETY AND INCIDENTS

SAFETY ISSUES THAT NEED TO BE ADDRESSED	ACCIDENTS / INCIDENTS / STEPS NEEDED TO RESOLVE

SUMMARY OF THE WORK DONE TODAY

_____
_____
_____
_____
_____

IMPORTANT NOTES

_____
_____
_____
_____
_____

NAME	SIGNATURE

TODAY LABOR

INITIALS	TRADE	START	FINISH	PAID HOURS	OVERTIME	COMPANY
☐ EMPLOYEE ☐ CONTRUCTOR		AM	PM			
☐ EMPLOYEE ☐ CONTRUCTOR		AM	PM			
☐ EMPLOYEE ☐ CONTRUCTOR		AM	PM			
☐ EMPLOYEE ☐ CONTRUCTOR		AM	PM			
☐ EMPLOYEE ☐ CONTRUCTOR		AM	PM			
☐ EMPLOYEE ☐ CONTRUCTOR		AM	PM			
☐ EMPLOYEE ☐ CONTRUCTOR		AM	PM			
☐ EMPLOYEE ☐ CONTRUCTOR		AM	PM			

EQUIPMENT ON SITE	NO. OF UNITE	WORKING YES / NO

HIRED EQUIPMENT	NO. OF UNITE	EQUIPMENT RENTED	FROM	RATE

NAME: _____   SIGNATURE: _____

MO   TU   WE   TH   FR   SA   SU
☐   ☐   ☐   ☐   ☐   ☐   ☐

DATE: ___ / ___ / ___

PROJECT:

FOREMAN:

WEATHER

F°____   C°____   ____ AM   ____ PM

HOURS DUE TO BAD WEATHER	ISSUED AND DELAYS

NOTE: _____

COMPLETION DATE	DAYS AHEAD OF SCHEDULE	DAYS BEHIND SCHEDULE

SAFETY AND INCIDENTS

SAFETY ISSUES THAT NEED TO BE ADDRESSED	ACCIDENTS / INCIDENTS / STEPS NEEDED TO RESOLVE

SUMMARY OF THE WORK DONE TODAY

_____
_____
_____
_____
_____

IMPORTANT NOTES

_____
_____
_____
_____
_____

NAME	SIGNATURE

TODAY LABOR

INITIALS	TRADE	START	FINISH	PAID HOURS	OVERTIME	COMPANY
☐ EMPLOYEE ☐ CONTRUCTOR		AM	PM			
☐ EMPLOYEE ☐ CONTRUCTOR		AM	PM			
☐ EMPLOYEE ☐ CONTRUCTOR		AM	PM			
☐ EMPLOYEE ☐ CONTRUCTOR		AM	PM			
☐ EMPLOYEE ☐ CONTRUCTOR		AM	PM			
☐ EMPLOYEE ☐ CONTRUCTOR		AM	PM			
☐ EMPLOYEE ☐ CONTRUCTOR		AM	PM			
☐ EMPLOYEE ☐ CONTRUCTOR		AM	PM			

EQUIPMENT ON SITE	NO. OF UNITE	WORKING YES / NO

HIRED EQUIPMENT	NO. OF UNITE	EQUIPMENT RENTED	FROM	RATE

NAME: _____   SIGNATURE: _____

MO TU WE TH FR SA SU

☐ ☐ ☐ ☐ ☐ ☐ ☐

DATE: / /

PROJECT:

FOREMAN:

WEATHER

F°_____ C°_____ _____AM _____PM

HOURS DUE TO BAD WEATHER

ISSUED AND DELAYS

NOTE:

COMPLETION DATE	DAYS AHEAD OF SCHEDULE	DAYS BEHIND SCHEDULE

SAFETY AND INCIDENTS

SAFETY ISSUES THAT NEED TO BE ADDRESSED	ACCIDENTS / INCIDENTS / STEPS NEEDED TO RESOLVE

SUMMARY OF THE WORK DONE TODAY

IMPORTANT NOTES

NAME	SIGNATURE

TODAY LABOR

INITIALS	TRADE	START	FINISH	PAID HOURS	OVERTIME	COMPANY
☐ EMPLOYEE ☐ CONTRUCTOR		AM	PM			
☐ EMPLOYEE ☐ CONTRUCTOR		AM	PM			
☐ EMPLOYEE ☐ CONTRUCTOR		AM	PM			
☐ EMPLOYEE ☐ CONTRUCTOR		AM	PM			
☐ EMPLOYEE ☐ CONTRUCTOR		AM	PM			
☐ EMPLOYEE ☐ CONTRUCTOR		AM	PM			
☐ EMPLOYEE ☐ CONTRUCTOR		AM	PM			
☐ EMPLOYEE ☐ CONTRUCTOR		AM	PM			

EQUIPMENT ON SITE	NO. OF UNITE	WORKING YES / NO

HIRED EQUIPMENT	NO. OF UNITE	EQUIPMENT RENTED	FROM	RATE

NAME: _____   SIGNATURE: _____

MO   TU   WE   TH   FR   SA   SU
☐   ☐   ☐   ☐   ☐   ☐   ☐   DATE: /   /

PROJECT: FOREMAN:

WEATHER | HOURS DUE TO | ISSUED AND DELAYS |
                                          | BAD WEATHER | |
         F°_____ C°_____ _____AM _____PM

NOTE: _____

COMPLETION DATE	DAYS AHEAD OF SCHEDULE	DAYS BEHIND SCHEDULE

SAFETY AND INCIDENTS

SAFETY ISSUES THAT NEED TO BE ADDRESSED	ACCIDENTS / INCIDENTS / STEPS NEEDED TO RESOLVE

SUMMARY OF THE WORK DONE TODAY

_____
_____
_____
_____
_____

IMPORTANT NOTES

_____
_____
_____
_____
_____

NAME	SIGNATURE

TODAY LABOR

INITIALS	TRADE	START	FINISH	PAID HOURS	OVERTIME	COMPANY
☐ EMPLOYEE ☐ CONTRUCTOR		AM	PM			
☐ EMPLOYEE ☐ CONTRUCTOR		AM	PM			
☐ EMPLOYEE ☐ CONTRUCTOR		AM	PM			
☐ EMPLOYEE ☐ CONTRUCTOR		AM	PM			
☐ EMPLOYEE ☐ CONTRUCTOR		AM	PM			
☐ EMPLOYEE ☐ CONTRUCTOR		AM	PM			
☐ EMPLOYEE ☐ CONTRUCTOR		AM	PM			
☐ EMPLOYEE ☐ CONTRUCTOR		AM	PM			

EQUIPMENT ON SITE	NO. OF UNITE	WORKING YES / NO

HIRED EQUIPMENT	NO. OF UNITE	EQUIPMENT RENTED	FROM	RATE

NAME: _____   SIGNATURE: _____

MO TU WE TH FR SA SU
☐   ☐   ☐   ☐   ☐   ☐   ☐

DATE: /   /

PROJECT:

FOREMAN:

WEATHER

F°_____ C°_____   _____ AM _____ PM

HOURS DUE TO BAD WEATHER	ISSUED AND DELAYS

NOTE: _____

COMPLETION DATE	DAYS AHEAD OF SCHEDULE	DAYS BEHIND SCHEDULE

SAFETY AND INCIDENTS

SAFETY ISSUES THAT NEED TO BE ADDRESSED	ACCIDENTS / INCIDENTS / STEPS NEEDED TO RESOLVE

SUMMARY OF THE WORK DONE TODAY

_____
_____
_____
_____
_____

IMPORTANT NOTES

_____
_____
_____
_____
_____

NAME	SIGNATURE

TODAY LABOR

INITIALS	TRADE	START	FINISH	PAID HOURS	OVERTIME	COMPANY
☐ EMPLOYEE ☐ CONTRUCTOR		AM	PM			
☐ EMPLOYEE ☐ CONTRUCTOR		AM	PM			
☐ EMPLOYEE ☐ CONTRUCTOR		AM	PM			
☐ EMPLOYEE ☐ CONTRUCTOR		AM	PM			
☐ EMPLOYEE ☐ CONTRUCTOR		AM	PM			
☐ EMPLOYEE ☐ CONTRUCTOR		AM	PM			
☐ EMPLOYEE ☐ CONTRUCTOR		AM	PM			
☐ EMPLOYEE ☐ CONTRUCTOR		AM	PM			

EQUIPMENT ON SITE	NO. OF UNITE	WORKING YES / NO

HIRED EQUIPMENT	NO. OF UNITE	EQUIPMENT RENTED	FROM	RATE

NAME: _____   SIGNATURE: _____

MO TU WE TH FR SA SU
☐   ☐   ☐   ☐   ☐   ☐   ☐

DATE: ___ / ___ / ___

PROJECT: _____

FOREMAN: _____

WEATHER ⛈ ⛅ ☁ 🌨 ☀ 🌧 ⛈

F° _____ C° _____ _____ AM _____ PM

HOURS DUE TO BAD WEATHER	ISSUED AND DELAYS

NOTE: _____

COMPLETION DATE	DAYS AHEAD OF SCHEDULE	DAYS BEHIND SCHEDULE

SAFETY AND INCIDENTS

SAFETY ISSUES THAT NEED TO BE ADDRESSED	ACCIDENTS / INCIDENTS / STEPS NEEDED TO RESOLVE

SUMMARY OF THE WORK DONE TODAY

_____
_____
_____
_____
_____
_____

IMPORTANT NOTES

_____
_____
_____
_____
_____
_____

NAME	SIGNATURE

TODAY LABOR

INITIALS	TRADE	START	FINISH	PAID HOURS	OVERTIME	COMPANY
☐ EMPLOYEE ☐ CONTRUCTOR		AM	PM			
☐ EMPLOYEE ☐ CONTRUCTOR		AM	PM			
☐ EMPLOYEE ☐ CONTRUCTOR		AM	PM			
☐ EMPLOYEE ☐ CONTRUCTOR		AM	PM			
☐ EMPLOYEE ☐ CONTRUCTOR		AM	PM			
☐ EMPLOYEE ☐ CONTRUCTOR		AM	PM			
☐ EMPLOYEE ☐ CONTRUCTOR		AM	PM			
☐ EMPLOYEE ☐ CONTRUCTOR		AM	PM			

EQUIPMENT ON SITE	NO. OF UNITE	WORKING YES / NO

HIRED EQUIPMENT	NO. OF UNITE	EQUIPMENT RENTED	FROM	RATE

NAME: _____   SIGNATURE: _____

MO TU WE TH FR SA SU
☐   ☐   ☐   ☐   ☐   ☐   ☐

DATE: /   /

PROJECT:

FOREMAN:

WEATHER

F°_____ C°_____   _____AM _____PM

HOURS DUE TO BAD WEATHER	ISSUED AND DELAYS

NOTE: _____

COMPLETION DATE	DAYS AHEAD OF SCHEDULE	DAYS BEHIND SCHEDULE

SAFETY AND INCIDENTS

SAFETY ISSUES THAT NEED TO BE ADDRESSED	ACCIDENTS / INCIDENTS / STEPS NEEDED TO RESOLVE

SUMMARY OF THE WORK DONE TODAY

IMPORTANT NOTES

NAME	SIGNATURE

TODAY LABOR

INITIALS	TRADE	START	FINISH	PAID HOURS	OVERTIME	COMPANY
☐ EMPLOYEE ☐ CONTRUCTOR		AM	PM			
☐ EMPLOYEE ☐ CONTRUCTOR		AM	PM			
☐ EMPLOYEE ☐ CONTRUCTOR		AM	PM			
☐ EMPLOYEE ☐ CONTRUCTOR		AM	PM			
☐ EMPLOYEE ☐ CONTRUCTOR		AM	PM			
☐ EMPLOYEE ☐ CONTRUCTOR		AM	PM			
☐ EMPLOYEE ☐ CONTRUCTOR		AM	PM			
☐ EMPLOYEE ☐ CONTRUCTOR		AM	PM			

EQUIPMENT ON SITE	NO. OF UNITE	WORKING YES / NO

HIRED EQUIPMENT	NO. OF UNITE	EQUIPMENT RENTED	FROM	RATE

NAME: _____   SIGNATURE: _____

MO TU WE TH FR SA SU
☐   ☐   ☐   ☐   ☐   ☐   ☐

DATE: ___ / ___ / ___

PROJECT: _____

FOREMAN: _____

WEATHER ⛅ 🌦 ☁ 🌨 ☀ 🌧 ⛈

F° _____ C° _____   _____ AM _____ PM

HOURS DUE TO BAD WEATHER	ISSUED AND DELAYS
	_____

NOTE: _____

COMPLETION DATE	DAYS AHEAD OF SCHEDULE	DAYS BEHIND SCHEDULE

SAFETY AND INCIDENTS

SAFETY ISSUES THAT NEED TO BE ADDRESSED	ACCIDENTS / INCIDENTS / STEPS NEEDED TO RESOLVE
_____	_____
_____	_____
_____	_____
_____	_____

SUMMARY OF THE WORK DONE TODAY

_____
_____
_____
_____
_____
_____

IMPORTANT NOTES

_____
_____
_____
_____
_____
_____

NAME	SIGNATURE

TODAY LABOR

INITIALS	TRADE	START	FINISH	PAID HOURS	OVERTIME	COMPANY
☐ EMPLOYEE ☐ CONTRUCTOR		AM	PM			
☐ EMPLOYEE ☐ CONTRUCTOR		AM	PM			
☐ EMPLOYEE ☐ CONTRUCTOR		AM	PM			
☐ EMPLOYEE ☐ CONTRUCTOR		AM	PM			
☐ EMPLOYEE ☐ CONTRUCTOR		AM	PM			
☐ EMPLOYEE ☐ CONTRUCTOR		AM	PM			
☐ EMPLOYEE ☐ CONTRUCTOR		AM	PM			
☐ EMPLOYEE ☐ CONTRUCTOR		AM	PM			

EQUIPMENT ON SITE	NO. OF UNITE	WORKING YES / NO

HIRED EQUIPMENT	NO. OF UNITE	EQUIPMENT RENTED	FROM	RATE

NAME: _____   SIGNATURE: _____

MO TU WE TH FR SA SU
☐ ☐ ☐ ☐ ☐ ☐ ☐

DATE: ___ / ___ / ___

PROJECT:

FOREMAN:

WEATHER F°_____ C°_____ _____AM _____PM

HOURS DUE TO BAD WEATHER	ISSUED AND DELAYS

NOTE: _____

COMPLETION DATE	DAYS AHEAD OF SCHEDULE	DAYS BEHIND SCHEDULE

SAFETY AND INCIDENTS

SAFETY ISSUES THAT NEED TO BE ADDRESSED	ACCIDENTS / INCIDENTS / STEPS NEEDED TO RESOLVE

SUMMARY OF THE WORK DONE TODAY

IMPORTANT NOTES

NAME	SIGNATURE

TODAY LABOR

INITIALS	TRADE	START	FINISH	PAID HOURS	OVERTIME	COMPANY
☐ EMPLOYEE ☐ CONTRUCTOR		AM	PM			
☐ EMPLOYEE ☐ CONTRUCTOR		AM	PM			
☐ EMPLOYEE ☐ CONTRUCTOR		AM	PM			
☐ EMPLOYEE ☐ CONTRUCTOR		AM	PM			
☐ EMPLOYEE ☐ CONTRUCTOR		AM	PM			
☐ EMPLOYEE ☐ CONTRUCTOR		AM	PM			
☐ EMPLOYEE ☐ CONTRUCTOR		AM	PM			
☐ EMPLOYEE ☐ CONTRUCTOR		AM	PM			

EQUIPMENT ON SITE	NO. OF UNITE	WORKING YES / NO

HIRED EQUIPMENT	NO. OF UNITE	EQUIPMENT RENTED	FROM	RATE

NAME: _____   SIGNATURE: _____

MO TU WE TH FR SA SU
☐   ☐   ☐   ☐   ☐   ☐   ☐

DATE: ___ / ___ / ___

PROJECT: _____

FOREMAN: _____

WEATHER ☁️ 🌤️ ☁️ 🌨️ ☀️ 🌧️ ⛈️

F°_____ C°_____   _____ AM _____ PM

HOURS DUE TO BAD WEATHER	ISSUED AND DELAYS

NOTE: _____

COMPLETION DATE	DAYS AHEAD OF SCHEDULE	DAYS BEHIND SCHEDULE

SAFETY AND INCIDENTS

SAFETY ISSUES THAT NEED TO BE ADDRESSED	ACCIDENTS / INCIDENTS / STEPS NEEDED TO RESOLVE

SUMMARY OF THE WORK DONE TODAY

_____
_____
_____
_____
_____
_____

IMPORTANT NOTES

_____
_____
_____
_____
_____

NAME	SIGNATURE

TODAY LABOR

INITIALS	TRADE	START	FINISH	PAID HOURS	OVERTIME	COMPANY
☐ EMPLOYEE ☐ CONTRUCTOR		AM	PM			
☐ EMPLOYEE ☐ CONTRUCTOR		AM	PM			
☐ EMPLOYEE ☐ CONTRUCTOR		AM	PM			
☐ EMPLOYEE ☐ CONTRUCTOR		AM	PM			
☐ EMPLOYEE ☐ CONTRUCTOR		AM	PM			
☐ EMPLOYEE ☐ CONTRUCTOR		AM	PM			
☐ EMPLOYEE ☐ CONTRUCTOR		AM	PM			
☐ EMPLOYEE ☐ CONTRUCTOR		AM	PM			

EQUIPMENT ON SITE	NO. OF UNITE	WORKING YES / NO

HIRED EQUIPMENT	NO. OF UNITE	EQUIPMENT RENTED	FROM	RATE

NAME: _____   SIGNATURE: _____

MO   TU   WE   TH   FR   SA   SU
☐   ☐   ☐   ☐   ☐   ☐   ☐

DATE: /   /

PROJECT:

FOREMAN:

WEATHER ☁ ⛅ ☁ 🌨 ☀ 🌧 ⛈

F°_____ C°_____   _____AM _____PM

HOURS DUE TO BAD WEATHER	ISSUED AND DELAYS

NOTE: _____

COMPLETION DATE	DAYS AHEAD OF SCHEDULE	DAYS BEHIND SCHEDULE

SAFETY AND INCIDENTS

SAFETY ISSUES THAT NEED TO BE ADDRESSED	ACCIDENTS / INCIDENTS / STEPS NEEDED TO RESOLVE

SUMMARY OF THE WORK DONE TODAY

_____
_____
_____
_____
_____

IMPORTANT NOTES

_____
_____
_____
_____
_____

NAME	SIGNATURE

TODAY LABOR

INITIALS	TRADE	START	FINISH	PAID HOURS	OVERTIME	COMPANY
☐ EMPLOYEE ☐ CONTRUCTOR		AM	PM			
☐ EMPLOYEE ☐ CONTRUCTOR		AM	PM			
☐ EMPLOYEE ☐ CONTRUCTOR		AM	PM			
☐ EMPLOYEE ☐ CONTRUCTOR		AM	PM			
☐ EMPLOYEE ☐ CONTRUCTOR		AM	PM			
☐ EMPLOYEE ☐ CONTRUCTOR		AM	PM			
☐ EMPLOYEE ☐ CONTRUCTOR		AM	PM			
☐ EMPLOYEE ☐ CONTRUCTOR		AM	PM			

EQUIPMENT ON SITE	NO. OF UNITE	WORKING YES / NO

HIRED EQUIPMENT	NO. OF UNITE	EQUIPMENT RENTED	FROM	RATE

NAME: _____   SIGNATURE: _____

MO TU WE TH FR SA SU
☐ ☐ ☐ ☐ ☐ ☐ ☐

DATE: ___ / ___ / ___

PROJECT:

FOREMAN:

WEATHER

F°_____ C°_____ _____ AM _____ PM

HOURS DUE TO BAD WEATHER	ISSUED AND DELAYS

NOTE: _____

COMPLETION DATE	DAYS AHEAD OF SCHEDULE	DAYS BEHIND SCHEDULE

SAFETY AND INCIDENTS

SAFETY ISSUES THAT NEED TO BE ADDRESSED	ACCIDENTS / INCIDENTS / STEPS NEEDED TO RESOLVE

SUMMARY OF THE WORK DONE TODAY

IMPORTANT NOTES

NAME	SIGNATURE

TODAY LABOR

INITIALS	TRADE	START	FINISH	PAID HOURS	OVERTIME	COMPANY
☐ EMPLOYEE ☐ CONTRUCTOR		AM	PM			
☐ EMPLOYEE ☐ CONTRUCTOR		AM	PM			
☐ EMPLOYEE ☐ CONTRUCTOR		AM	PM			
☐ EMPLOYEE ☐ CONTRUCTOR		AM	PM			
☐ EMPLOYEE ☐ CONTRUCTOR		AM	PM			
☐ EMPLOYEE ☐ CONTRUCTOR		AM	PM			
☐ EMPLOYEE ☐ CONTRUCTOR		AM	PM			
☐ EMPLOYEE ☐ CONTRUCTOR		AM	PM			

EQUIPMENT ON SITE	NO. OF UNITE	WORKING YES / NO

HIRED EQUIPMENT	NO. OF UNITE	EQUIPMENT RENTED	FROM	RATE

NAME: _____   SIGNATURE: _____

MO   TU   WE   TH   FR   SA   SU   DATE: /   /
☐   ☐   ☐   ☐   ☐   ☐   ☐

PROJECT: FOREMAN:

WEATHER   ☁☂ ⛅ ☁ 🌨 ☀ 🌧 ⛈ | HOURS DUE TO | ISSUED AND DELAYS
                                              | BAD WEATHER |
           F°_____ C°_____ _____AM _____PM | |

NOTE: _____

COMPLETION DATE	DAYS AHEAD OF SCHEDULE	DAYS BEHIND SCHEDULE

SAFETY AND INCIDENTS

SAFETY ISSUES THAT NEED TO BE ADDRESSED	ACCIDENTS / INCIDENTS / STEPS NEEDED TO RESOLVE
_____	_____
_____	_____
_____	_____
_____	_____

SUMMARY OF THE WORK DONE TODAY

_____
_____
_____
_____
_____
_____

IMPORTANT NOTES

_____
_____
_____
_____
_____
_____

NAME	SIGNATURE

TODAY LABOR

INITIALS	TRADE	START	FINISH	PAID HOURS	OVERTIME	COMPANY
☐ EMPLOYEE ☐ CONTRUCTOR		AM	PM			
☐ EMPLOYEE ☐ CONTRUCTOR		AM	PM			
☐ EMPLOYEE ☐ CONTRUCTOR		AM	PM			
☐ EMPLOYEE ☐ CONTRUCTOR		AM	PM			
☐ EMPLOYEE ☐ CONTRUCTOR		AM	PM			
☐ EMPLOYEE ☐ CONTRUCTOR		AM	PM			
☐ EMPLOYEE ☐ CONTRUCTOR		AM	PM			
☐ EMPLOYEE ☐ CONTRUCTOR		AM	PM			

EQUIPMENT ON SITE	NO. OF UNITE	WORKING YES / NO

HIRED EQUIPMENT	NO. OF UNITE	EQUIPMENT RENTED	FROM	RATE

NAME: _____   SIGNATURE: _____

MO TU WE TH FR SA SU
☐   ☐   ☐   ☐   ☐   ☐   ☐

DATE: ___/___/___

PROJECT: _____

FOREMAN: _____

WEATHER ☁ ⛅ ☁ 🌨 ☀ 🌧 ⛈

F°____ C°____ ____AM ____PM

| HOURS DUE TO BAD WEATHER | ISSUED AND DELAYS |

NOTE: _____

COMPLETION DATE	DAYS AHEAD OF SCHEDULE	DAYS BEHIND SCHEDULE

SAFETY AND INCIDENTS

SAFETY ISSUES THAT NEED TO BE ADDRESSED	ACCIDENTS / INCIDENTS / STEPS NEEDED TO RESOLVE
_____	_____
_____	_____
_____	_____
_____	_____

SUMMARY OF THE WORK DONE TODAY

_____
_____
_____
_____
_____
_____

IMPORTANT NOTES

_____
_____
_____
_____
_____
_____
_____

NAME	SIGNATURE

TODAY LABOR

INITIALS	TRADE	START	FINISH	PAID HOURS	OVERTIME	COMPANY
☐ EMPLOYEE ☐ CONTRUCTOR		AM	PM			
☐ EMPLOYEE ☐ CONTRUCTOR		AM	PM			
☐ EMPLOYEE ☐ CONTRUCTOR		AM	PM			
☐ EMPLOYEE ☐ CONTRUCTOR		AM	PM			
☐ EMPLOYEE ☐ CONTRUCTOR		AM	PM			
☐ EMPLOYEE ☐ CONTRUCTOR		AM	PM			
☐ EMPLOYEE ☐ CONTRUCTOR		AM	PM			
☐ EMPLOYEE ☐ CONTRUCTOR		AM	PM			

EQUIPMENT ON SITE	NO. OF UNITE	WORKING YES / NO

HIRED EQUIPMENT	NO. OF UNITE	EQUIPMENT RENTED	FROM	RATE

NAME: _____   SIGNATURE: _____

MO	TU	WE	TH	FR	SA	SU
☐	☐	☐	☐	☐	☐	☐

DATE: / /

PROJECT:

FOREMAN:

WEATHER

F°_____ C°_____ _____ AM _____ PM

HOURS DUE TO BAD WEATHER	ISSUED AND DELAYS

NOTE: _____

COMPLETION DATE	DAYS AHEAD OF SCHEDULE	DAYS BEHIND SCHEDULE

SAFETY AND INCIDENTS

SAFETY ISSUES THAT NEED TO BE ADDRESSED	ACCIDENTS / INCIDENTS / STEPS NEEDED TO RESOLVE

SUMMARY OF THE WORK DONE TODAY

IMPORTANT NOTES

NAME	SIGNATURE

TODAY LABOR

INITIALS	TRADE	START	FINISH	PAID HOURS	OVERTIME	COMPANY
☐ EMPLOYEE ☐ CONTRUCTOR		AM	PM			
☐ EMPLOYEE ☐ CONTRUCTOR		AM	PM			
☐ EMPLOYEE ☐ CONTRUCTOR		AM	PM			
☐ EMPLOYEE ☐ CONTRUCTOR		AM	PM			
☐ EMPLOYEE ☐ CONTRUCTOR		AM	PM			
☐ EMPLOYEE ☐ CONTRUCTOR		AM	PM			
☐ EMPLOYEE ☐ CONTRUCTOR		AM	PM			
☐ EMPLOYEE ☐ CONTRUCTOR		AM	PM			

EQUIPMENT ON SITE	NO. OF UNITE	WORKING YES / NO

HIRED EQUIPMENT	NO. OF UNITE	EQUIPMENT RENTED	FROM	RATE

NAME: _____   SIGNATURE: _____

MO TU WE TH FR SA SU ☐ ☐ ☐ ☐ ☐ ☐ ☐

DATE: / /

PROJECT:

FOREMAN:

WEATHER

F°_____ C°_____ _____AM _____PM

HOURS DUE TO BAD WEATHER

ISSUED AND DELAYS

NOTE: _____

COMPLETION DATE	DAYS AHEAD OF SCHEDULE	DAYS BEHIND SCHEDULE

SAFETY AND INCIDENTS

SAFETY ISSUES THAT NEED TO BE ADDRESSED	ACCIDENTS / INCIDENTS / STEPS NEEDED TO RESOLVE

SUMMARY OF THE WORK DONE TODAY

IMPORTANT NOTES

NAME	SIGNATURE

TODAY LABOR

INITIALS	TRADE	START	FINISH	PAID HOURS	OVERTIME	COMPANY
☐ EMPLOYEE ☐ CONTRUCTOR		AM	PM			
☐ EMPLOYEE ☐ CONTRUCTOR		AM	PM			
☐ EMPLOYEE ☐ CONTRUCTOR		AM	PM			
☐ EMPLOYEE ☐ CONTRUCTOR		AM	PM			
☐ EMPLOYEE ☐ CONTRUCTOR		AM	PM			
☐ EMPLOYEE ☐ CONTRUCTOR		AM	PM			
☐ EMPLOYEE ☐ CONTRUCTOR		AM	PM			
☐ EMPLOYEE ☐ CONTRUCTOR		AM	PM			

EQUIPMENT ON SITE	NO. OF UNITE	WORKING YES / NO

HIRED EQUIPMENT	NO. OF UNITE	EQUIPMENT RENTED	FROM	RATE

NAME: _____   SIGNATURE: _____

MO TU WE TH FR SA SU
☐   ☐   ☐   ☐   ☐   ☐   ☐

DATE: ___ / ___ / ___

PROJECT:

FOREMAN:

WEATHER

F° _____ C° _____ _____ AM _____ PM

HOURS DUE TO BAD WEATHER	ISSUED AND DELAYS

COMPLETION DATE	DAYS AHEAD OF SCHEDULE	DAYS BEHIND SCHEDULE

SAFETY AND INCIDENTS

SAFETY ISSUES THAT NEED TO BE ADDRESSED	ACCIDENTS / INCIDENTS / STEPS NEEDED TO RESOLVE

SUMMARY OF THE WORK DONE TODAY

IMPORTANT NOTES

NAME	SIGNATURE

TODAY LABOR

INITIALS	TRADE	START	FINISH	PAID HOURS	OVERTIME	COMPANY
☐ EMPLOYEE ☐ CONTRUCTOR		AM	PM			
☐ EMPLOYEE ☐ CONTRUCTOR		AM	PM			
☐ EMPLOYEE ☐ CONTRUCTOR		AM	PM			
☐ EMPLOYEE ☐ CONTRUCTOR		AM	PM			
☐ EMPLOYEE ☐ CONTRUCTOR		AM	PM			
☐ EMPLOYEE ☐ CONTRUCTOR		AM	PM			
☐ EMPLOYEE ☐ CONTRUCTOR		AM	PM			
☐ EMPLOYEE ☐ CONTRUCTOR		AM	PM			

EQUIPMENT ON SITE	NO. OF UNITE	WORKING YES / NO

HIRED EQUIPMENT	NO. OF UNITE	EQUIPMENT RENTED	FROM	RATE

NAME: _____   SIGNATURE: _____

MO TU WE TH FR SA SU
☐   ☐   ☐   ☐   ☐   ☐   ☐

DATE: ___ / ___ / ___

PROJECT:

FOREMAN:

WEATHER

F°_____ C°_____   _____ AM _____ PM

HOURS DUE TO
BAD WEATHER

ISSUED AND DELAYS

NOTE: _____

COMPLETION DATE	DAYS AHEAD OF SCHEDULE	DAYS BEHIND SCHEDULE

SAFETY AND INCIDENTS

SAFETY ISSUES THAT NEED TO BE ADDRESSED	ACCIDENTS / INCIDENTS / STEPS NEEDED TO RESOLVE

SUMMARY OF THE WORK DONE TODAY

IMPORTANT NOTES

NAME	SIGNATURE

TODAY LABOR

INITIALS	TRADE	START	FINISH	PAID HOURS	OVERTIME	COMPANY
☐ EMPLOYEE ☐ CONTRUCTOR		AM	PM			
☐ EMPLOYEE ☐ CONTRUCTOR		AM	PM			
☐ EMPLOYEE ☐ CONTRUCTOR		AM	PM			
☐ EMPLOYEE ☐ CONTRUCTOR		AM	PM			
☐ EMPLOYEE ☐ CONTRUCTOR		AM	PM			
☐ EMPLOYEE ☐ CONTRUCTOR		AM	PM			
☐ EMPLOYEE ☐ CONTRUCTOR		AM	PM			
☐ EMPLOYEE ☐ CONTRUCTOR		AM	PM			

EQUIPMENT ON SITE	NO. OF UNITE	WORKING YES / NO

HIRED EQUIPMENT	NO. OF UNITE	EQUIPMENT RENTED	FROM	RATE

NAME: _____   SIGNATURE: _____

MO TU WE TH FR SA SU
☐   ☐   ☐   ☐   ☐   ☐   ☐

DATE: ___/___/___

PROJECT: _____

FOREMAN: _____

WEATHER

F°____ C°____   ____AM ____PM

HOURS DUE TO BAD WEATHER	ISSUED AND DELAYS

NOTE: _____

COMPLETION DATE	DAYS AHEAD OF SCHEDULE	DAYS BEHIND SCHEDULE

SAFETY AND INCIDENTS

SAFETY ISSUES THAT NEED TO BE ADDRESSED	ACCIDENTS / INCIDENTS / STEPS NEEDED TO RESOLVE

SUMMARY OF THE WORK DONE TODAY

_____
_____
_____
_____
_____

IMPORTANT NOTES

_____
_____
_____
_____
_____

NAME	SIGNATURE

TODAY LABOR

INITIALS	TRADE	START	FINISH	PAID HOURS	OVERTIME	COMPANY
☐ EMPLOYEE ☐ CONTRUCTOR		AM	PM			
☐ EMPLOYEE ☐ CONTRUCTOR		AM	PM			
☐ EMPLOYEE ☐ CONTRUCTOR		AM	PM			
☐ EMPLOYEE ☐ CONTRUCTOR		AM	PM			
☐ EMPLOYEE ☐ CONTRUCTOR		AM	PM			
☐ EMPLOYEE ☐ CONTRUCTOR		AM	PM			
☐ EMPLOYEE ☐ CONTRUCTOR		AM	PM			
☐ EMPLOYEE ☐ CONTRUCTOR		AM	PM			

EQUIPMENT ON SITE	NO. OF UNITE	WORKING YES / NO

HIRED EQUIPMENT	NO. OF UNITE	EQUIPMENT RENTED	FROM	RATE

NAME: _____   SIGNATURE: _____

MO   TU   WE   TH   FR   SA   SU
☐   ☐   ☐   ☐   ☐   ☐   ☐

DATE: /   /

PROJECT:

FOREMAN:

WEATHER

F°_____   C°_____   _____ AM   _____ PM

HOURS DUE TO BAD WEATHER	ISSUED AND DELAYS

NOTE: _____

COMPLETION DATE	DAYS AHEAD OF SCHEDULE	DAYS BEHIND SCHEDULE

SAFETY AND INCIDENTS

SAFETY ISSUES THAT NEED TO BE ADDRESSED	ACCIDENTS / INCIDENTS / STEPS NEEDED TO RESOLVE

SUMMARY OF THE WORK DONE TODAY

IMPORTANT NOTES

NAME	SIGNATURE

TODAY LABOR

INITIALS	TRADE	START	FINISH	PAID HOURS	OVERTIME	COMPANY
☐ EMPLOYEE ☐ CONTRUCTOR		AM	PM			
☐ EMPLOYEE ☐ CONTRUCTOR		AM	PM			
☐ EMPLOYEE ☐ CONTRUCTOR		AM	PM			
☐ EMPLOYEE ☐ CONTRUCTOR		AM	PM			
☐ EMPLOYEE ☐ CONTRUCTOR		AM	PM			
☐ EMPLOYEE ☐ CONTRUCTOR		AM	PM			
☐ EMPLOYEE ☐ CONTRUCTOR		AM	PM			
☐ EMPLOYEE ☐ CONTRUCTOR		AM	PM			

EQUIPMENT ON SITE	NO. OF UNITE	WORKING YES / NO

HIRED EQUIPMENT	NO. OF UNITE	EQUIPMENT RENTED	FROM	RATE

NAME: _____   SIGNATURE: _____

IMPORTANT TELEPHONE NUMBER

▶ NAME _____ PHONE _____

EMAIL _____

▶ NAME _____ PHONE _____

EMAIL _____

▶ NAME _____ PHONE _____

EMAIL _____

▶ NAME _____ PHONE _____

EMAIL _____

▶ NAME _____ PHONE _____

EMAIL _____

▶ NAME _____ PHONE _____

EMAIL _____

▶ NAME _____ PHONE _____

EMAIL _____

▶ NAME _____ PHONE _____

EMAIL _____

▶ NAME _____ PHONE _____

EMAIL _____

▶ NAME _____ PHONE _____

EMAIL _____

▶ NAME _____ PHONE _____

EMAIL _____

▶ NAME _____ PHONE _____

EMAIL _____

▶ NAME _____ PHONE _____

EMAIL _____

www.ingramcontent.com/pod-product-compliance
Lightning Source LLC
Chambersburg PA
CBHW051758200326
41597CB00025B/4608